Mathshree By Himanshri

Copyright Page

Mathshree By Himanshri

ISBN: Paperback / Softback: 978-93-344-1228-4

Hardback: 978-93-5655-128-2

Ebook: 978-93-5626-425-0

Publisher Information: Himanshri Chaudhary ; Address: Ajeet Nagar, Jattari, Aligarh- 202137, INDIA

Author Details: Himanshri Chaudhary

Website / Contact Information:

Email: chimanshri143@gmail.com

Disclaimer:

Edition Information: First Edition

Cover Design & Interior Design Credits: Paremvir Malik

Foreword

Have you ever looked at a string of numbers and felt your brain freeze? Have you ever marveled at someone who can solve complex problems in their head with effortless speed? For most of us, math feels like a foreign language—a cryptic code reserved for a select few. This book is a radical act of rebellion against that idea. It is not a textbook filled with dry theory, but a playbook for a new kind of game, one where you are the master player.

Inside these pages, you will learn to see numbers not as enemies, but as allies. You will discover simple, elegant shortcuts that transform intimidating equations into fun challenges. The goal isn't just to get the right answer; it's to build a more agile, confident mind. By the time you reach the final chapter, you'll possess a new superpower: the ability to perform "mathemagic" with poise and precision. The journey starts now.

Preface

Mathematics is a vital part of academics, competitive exams, and daily life. Many students find calculations slow and difficult, which can affect their confidence. This book is created to simplify and speed up mathematical problem-solving.

Drawing from my teaching experience, I have gathered practical shortcuts anyone can learn and use. The book aims to help not only students but also competitive exam aspirants and teachers make math easier and more enjoyable.

I hope readers will gain speed, accuracy, and a deeper interest in math by practicing these tricks regularly.

Acknowledgement

I would like to express my heartfelt gratitude to all those who have supported and encouraged me throughout the journey of writing this book on short tricks for mathematical calculations.

I would like to sincerely thank **Mr. Paremvir Malik** for his invaluable encouragement and support in inspiring me to write this book. His motivation and guidance have been instrumental in my journey to create a resource that will be helpful for students, competitive exam aspirants, and teachers alike. I am deeply grateful for his belief in this project and for encouraging me to share these mathematical short tricks to make learning easier and faster for many.

A very special note of thanks goes to my sister, **Varsha Chaudhary**. I am incredibly grateful for her dedication in cross-checking the entire manuscript. She personally verified every mathematical trick and calculation in this book to ensure accuracy and clarity. Her sharp eye and commitment to excellence have been vital in making this book a reliable tool for learners.

I am also grateful to my family for their unwavering support and patience during the writing process. Thank you all for being an essential part of this endeavor.

WHY THIS BOOK...?

The main reason I wrote this book is to provide an easy-to-use resource that teaches smart math shortcuts. Traditional methods can be slow and daunting, especially during exams.

This book empowers learners and teachers alike by making math quicker and less stressful. My goal is to help students and aspirants solve problems faster without losing accuracy.

By using this book, I want readers to feel more confident and engaged in their math journey.

Table of Contents

MEET YOUR NEW
MATH SUPERPOWERS

Chapter-1

Meet Your New Math Superpowers

A New Way to See Math

Are you one of the many people who get a little nervous when you see a long line of numbers? Does the idea of doing math in your head feel impossible? You're not alone! Many people feel intimidated by arithmetic. It can seem like a long, boring process of steps and memorization. But what if there was another way?

This book isn't about memorizing endless rules. It's about unlocking the secrets that make numbers work. Think of it less like a school lesson and more like a secret training manual for your brain. By the time you're done, you'll have a set of "math superpowers" that will allow you to solve problems faster and more confidently than you ever thought possible.

What is a "Math Shortcut"?

A shortcut isn't a cheat. It's a clever trick or a different way of looking at a problem that gets you to the answer much more quickly. It's about thinking smarter, not harder. For example, did you know you can multiply any number by 11 in a matter of seconds, all in your head? Or that you can find the answer to a tricky subtraction problem by turning it into a simple addition? This book is full of these kinds of "aha!" moments.

Why These Shortcuts Matter?

Learning these tricks will do more than just make you faster at math. You'll start to see patterns in numbers that you never noticed before. This new way of thinking will boost your confidence and make you feel a sense of accomplishment every time you solve a problem. Whether you're a student trying to ace a test, an adult who wants to be quicker with numbers at work, or just someone who wants to exercise your brain, these shortcuts are for you.

So, let's get started. Turn the page, and get ready to discover the magic of numbers!

THE EASIEST WAY TO ADD AND SUBTRACT

Chapter-2

The Easiest Way to Add and Subtract

A Quick Note Before We Start

Before we dive into multiplication and division, let's master the most basic operations: addition and subtraction. While these might seem simple, you're about to learn how to do them faster and in your head. The key is to stop thinking like a calculator and start thinking like a human brain. We naturally think in big chunks, so we'll use that to our advantage.

The Power of Tens: The Rounding Trick

Sometimes, adding two numbers can be tricky if they don't end in zero. But what if you made one of them a round, easy number?

Let's try to solve 48 + 23.

The normal way involves lining up the numbers and adding the digits. Our shortcut involves making 48 easier to work with.

- **Step 1:** Round 48 up to the nearest ten, which is 50. How many did you add to get there? You added 2.
- **Step 2:** Now, do the easy addition: 50 + 23 = 73.
- **Step 3:** Remember that extra 2 you added? Now, you have to take it away. 73 - 2 = 71.

  And just like that, you have the answer: **71**.

  This works with subtraction too!

To **solve 95 - 28**, you can round 28 up to 30, which is a much easier number to subtract.

- **Step 1:** Round 28 up to 30. You added 2.
- **Step 2:** Do the easy subtraction: 95 - 30 = 65.
- **Step 3:** This time, because you subtracted *more* than you needed to, you must add that extra 2 back. 65 + 2 = 67.

The answer is **67**!

Left-to-Right Addition

In school, we learn to add from right to left (the ones place first). But our brains often process information from left to right. This trick uses that natural flow.

Let's solve 35 + 42.

- **Step 1:** Look at the numbers in the tens place first: 30 + 40. What does that equal? 70.

- **Step 2:** Now look at the numbers in the ones place: 5 + 2. That equals 7.
- **Step 3:** Add the two results together: 70 + 7 = 77.

You can do this with even bigger numbers.

Try 215 + 364.

- 200 + 300 = 500
- 10 + 60 = 70
- 5 + 4 = 9
- 500 + 70 + 9 = 579

This method helps you build the answer piece by piece in your head, just like you would say it out loud. It's a great way to improve your mental math.

Use the tricks you've learned to solve these in your head:

- 383+245=
- 238+162=
- 127+393=
- 810+238=
- 730+435=
- 450+210=

THE SECRETS OF SUBTRACTION
7
4
2
9
- 4 - 9 = 5
LEFT-TO-RIGHT HACK

Chapter – 3

The Subtraction Secrets

In the last chapter, we saw how you could turn tricky addition problems into easy ones. Now, let's do the same for subtraction. Many people find subtraction difficult because of "borrowing," but these tricks can help you skip that step entirely and get the answer in a flash!

The "All from Nine, Last from Ten" Trick

This is one of the coolest tricks in math! It's perfect for when you need to subtract a number from a number that is a power of 10, like 100, 1,000, 10,000, and so on.

Let's solve 1000−345:

Forget about borrowing! You'll solve this from left to right, one number at a time.

- **Step 1:** Look at the first digit of the number you're subtracting (345). Subtract it from 9: 9−3=6. This is the first digit of your answer.
- **Step 2:** Look at the next digit (4). Subtract it from 9: 9−4=5. This is the second digit of your answer.
- **Step 3:** Look at the last digit (5). Subtract it from 10, not 9: 10−5=5. This is the final digit of your answer.

Put it all together: 6, 5, 5. The answer is **655**.

This trick works every time for numbers like 100, 1,000, or any number that starts with a 1 and is followed by zeros.

The "Counting Up" Method:

This trick turns subtraction into a fun game of addition. It's especially useful for when the numbers are close together.

Let's solve 72−48:

Instead of subtracting, let's "count up" from the smaller number (48) to the bigger number (72) and see how much we added.

- **Step 1:** Start at 48. What's the closest round number? 50. To get from 48 to 50, you add 2.
- **Step 2:** Now, from 50, how much do you need to add to get to 72? You can add 22.
- **Step 3:** Add up the two numbers you just added: 2+22=24.

The answer is **24**.

Let's solve 278−134:

Instead of subtracting, let's "count up" from the smaller number (134) to the bigger number (278) and see how much we added.

- **Step 1:** Start at 134. What's the closest round number? 150. To get from 134 to 150, you add 16.
- **Step 2:** Now, from 150, how much do you need to add to get to 278? You can add 128.
- **Step 3:** Add up the two numbers you just added: 16+128=144.

The answer is **144**.

You solved a subtraction problem by only adding. This is a very powerful way to do subtraction in your head.

Try It Out!

Use the tricks you've learned to solve these in your head:

- 100−73=
- 900−456=
- 83−59=
- 51−27=
- 730-250=
- 278-134=
- 839-462=
- 94-37=
- 10000-8486=

5 TO 10 SECONDS MULTIPLICATIONTRICKS

Chapter-4

The 5 to 10 Seconds Multiplication Tricks

Now we're getting into the fun stuff! Multiplication can seem intimidating, especially when you can't use a calculator. But with these quick tricks, you'll be able to solve some common problems faster than you can type them into a machine.

Multiplying by 5: The "Halve and Shift" Trick

Multiplying by 5 is actually half of multiplying by 10. This simple trick uses that fact.

Let's solve 64×5:

- **Step 1:** Find half of the number: Half of 64 is 32.
- **Step 2:** Add a zero to the end of your answer: 320.

The answer is **320.**

This works for any number, whether it's even or odd!

If it's odd, you'll get a decimal. Just drop the decimal and add a 5 at the end. For example, 35×5 would be half of 35 (which is 17.5) with the decimal dropped and a 5 added on to make 175.

Multiplying by 9: The "Subtract One and Add" Trick

This one is a fun way to use subtraction for multiplication.

Let's solve 15×9.

- **Step 1:** Subtract 1 from the number you are multiplying by 9: 15−1=14. This is the first part of your answer.
- **Step 2:** Subtract the first part of your answer (14) from 15: 15−14=1. Now,

place this at the end of the number you got in step 1.

The answer is **135**!

(A more accurate way to explain it is:

```
15 * 9 = 15 * (10-1) = 150-15 = 135)
```

Or use the "last from ten, rest from nine" rule.

Let's try 15×9.

- **Step 1:** 15−1=14. That's the first two digits of your answer.
- **Step 2:** What do you add to 4 to make 9? 5. So the last digit of your answer is 5.

The answer is **135**! This trick works with any number!

The 11's Trick: The "Add the Neighbors" Rule

This is one of the coolest and easiest tricks to learn. It works for multiplying any two-digit number by 11.

Let's solve 45×11.

- **Step 1:** Take the number you are multiplying by 11 (). Split the digits apart and leave a space in the middle: 4_5.
- **Step 2:** Add the two digits together: 4+5=9.
- **Step 3:** Place that sum in the middle of the digits: 495.

That's it! The answer is **495**.

What happens if the sum is 10 or greater? You just "carry over" the 1, just like in normal addition.

Let's try 87×11.

- **Step 1:** Split the digits: 8_7.
- **Step 2:** Add the digits: 8+7=15.
- **Step 3:** Put the 5 in the middle and add the 1 to the first digit (8+1=9).

The result is **957.**

Try It Out!

Use the tricks you've learned to solve these in your head:

- $27 \times 11 = ?$

- $59 \times 11 = ?$

- $42 \times 5 = ?$

- $9 \times 9 = ?$

- $88 \times 9 = ?$

- $73 \times 5 = ?$

MULTIPLICATION MASTERS
THE MULTIPLLYING MIND

Chapter 5

Multiplication Masters

You've learned some excellent tricks for simple multiplication. Now, let's take your skills to the next level with some more advanced techniques that work on bigger numbers. These tricks are perfect for problems that look complicated but are surprisingly easy with the right approach.

The "Close to 100" Trick

This trick is a lifesaver for multiplying numbers that are just under 100, like 98×97. Instead of doing a long multiplication, we'll use 100 as our reference point.

Let's solve 98×97.

Step 1: Find the difference between each number and 100.

100−98=2

100−97=3

Step 2: Get the first part of your answer. Subtract one number's difference from the *other* number. It doesn't matter which one you choose, the answer will be the same!

98−3=95

(Or 97−2=95) So, the first part of our answer is 95.

Step 3: Get the last part of your answer. Multiply the two differences you found in Step 1.

2×3=6.

Since we are multiplying two numbers close to 100, we need a two-digit number here. So, we'll write this as 06.

Step 4: Put the two parts together.

95 and 06 become 9506.

The answer is **9,506**!

This trick also works for numbers close to 1,000, 10,000, and so on.

The "Splitting and Merging" Trick

This is a fun way to use what you already know (addition and simple multiplication) to solve a harder problem. It's based on a property called the distributive property.

Let's solve 15×12.

You might not know your 12 times table by heart, but you probably know your 10s and 2s. We can split 12 into 10+2.

Step 1: Break the problem into two parts using a number that's easy to multiply by, like 10.

15×10

15×2

Step 2: Solve each part.

$15 \times 10 = 150$

$15 \times 2 = 30$

Step 3: Add the results together!

$150 + 30 = 180$

The answer is **180**.

You can use this trick for any multiplication problem where one of the numbers can be broken down into easier parts.

Try It Out!

Use these tricks to solve the following problems in your head:

99×96=?

- 105×102=?

- 21×13=?

- 88×78=?

- 89×95=?

- 108×112=?

DIVISIBILITY RULES YOU'LL ACTUALLY USE

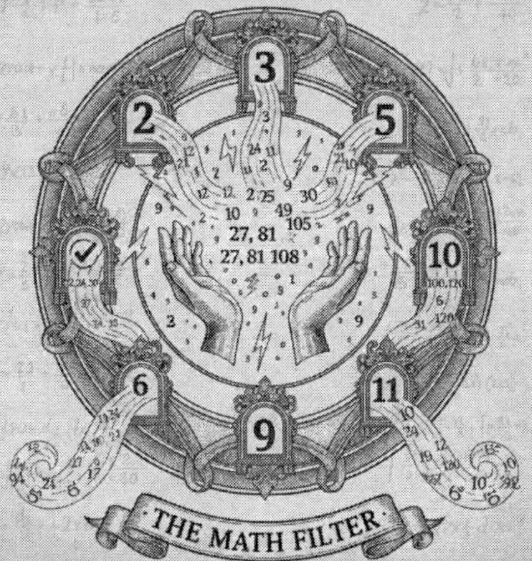

Chapter – 6

Divisibility Rules : You'll Actually Use

Ever wonder if a big, scary number can be divided by a smaller one without leaving a remainder? Instead of pulling out a calculator or doing long division, you can just use these simple tricks. Divisibility rules are like secret codes that tell you if one number fits perfectly into another.

Rule for 2

This is the easiest one! A number is divisible by 2 if it's an even number. This means it must end in a 0, 2, 4, 6, or 8.

- Example: Is 458 divisible by 2? Yes, because it ends in an 8.

Rule for 3

This one is a little bit like a puzzle. A number is divisible by 3 if the sum of its digits is divisible by 3.

- Example: Is 123 divisible by 3? Let's add the digits: 1+2+3=6. Since 6 is divisible by 3, the number 123 is also divisible by 3.

Rule for 4

A number is divisible by 4 if the number formed by its last two digits is divisible by 4.

- Example: Is 1,424 divisible by 4? Just look at the last two digits, 24. Since 24 is divisible by 4 (24÷4=6), the whole number is divisible by 4.

Rule for 5

This is another super-easy one. A number is divisible by 5 if it ends in a 0 or a 5.

- Example: Is 7,985 divisible by 5? Yes, because it ends in a 5.

Rule for 6

A number is divisible by 6 if it is divisible by both 2 and 3.

- Example: Is 36 divisible by 6? Yes, because it's even (divisible by 2) and the sum of its digits (3+6=9) is divisible by 3.

Rule for 8

A number is divisible by 8 if the number formed by its last three digits is divisible by 8. (This is most useful for large numbers.)

- Example: Is 1,520 divisible by 8? Look at the last three digits, 520. Since,

520÷8=65, the whole number is divisible by 8.

Rule for 9

This rule is very similar to the rule for 3. A number is divisible by 9 if the sum of its digits is divisible by 9.

- Example: Is 8,118 divisible by 9? The sum of the digits is 8+1+1+8=18. Since 18 is divisible by 9, the number is also divisible by 9.

Rule for 10

A number is divisible by 10 if it ends in a 0.

- Example: Is 5,670 divisible by 10? Yes, because it ends in a 0.

Try It Out

See if you can figure out which of these numbers are divisible by 3, 5, or 10 without doing any division.

- 125
- 68
- 273
- 550
- 11800
- 938
- 6432
- 891

SQUARING LIKE A GENIUS

Chapter – 7

Squaring Like a Genius

Finding the square of a number can seem like a lot of work. But what if there was a way to do it without multiplying the long way? In this chapter, you'll learn a couple of very simple patterns that will let you square numbers in your head, making you look like a total math genius.

Squaring Numbers That End in 5

This is one of the most famous and satisfying math tricks. It works for any two-digit number that ends in 5.

Let's find the answer to :

- **Step 1:** Look at the first digit of the number $(35)^2$. It's 3.
- **Step 2:** Multiply that first digit by the number that comes right after it. In this case, the number after 3 is 4. So,

3 times 4 =12. This is the first part of your answer.

- **Step 3:** Now for the easiest part. Just add 25 to the end of the number you got in Step 2.

Put it all together: 12 and 25 become 1,225.

That's it! This trick works every single time.

Try it with 65^2

The first digit is 6, so you multiply 6times7=42, and then you just add 25 to the end. The answer is **4,225**.

General Two-Digit Squaring

What if the number doesn't end in 5? Don't worry, there's a simple method for any two-digit number. We'll use a slightly different approach, but it's just as clever.

Let's solve 43^2.

- **Step 1:** Square the first digit: 4^2=16.

- **Step 2:** Square the second digit: $3^2=9$. Write it as 09.
- **Step 3:** Multiply the two digits and double it: 4 times 3 times 2 =24. Write this number starting one space from the right.

1609

+240

1849

The answer is **1,849**!

Try It Out!

Use these tricks to solve the following problems in your head:

- 25^2=?

- 85^2=?

- 21^2=?

- 54^2=?

- 38^2=?

- $97^2=?$

- $45^2=?$

- $63^2=?$

FINDING ROOTS WITHOUT CALCULATOR

Chapter – 8

Finding Roots Without a Calculator

After learning how to square numbers, it's time to learn how to do the opposite—find the square root! This might sound impossible without a calculator, but for perfect squares, it's surprisingly easy. You'll just need to use a little bit of detective work to solve the mystery.

The Two-Step Secret

This trick works by breaking the problem into two simple parts: finding the first digit of the answer and finding the last digit.

Let's find the square root of a perfect square like $\sqrt{1849}$

Step 1: Find the Last Digit

The last digit of the number you're trying to find the square root of tells you what the last digit of your answer could be.

- If the number ends in 1, the square root ends in a 1 or a 9.
- If the number ends in 4, the square root ends in a 2 or an 8.
- If the number ends in 9, the square root ends in a 3 or a 7.
- If the number ends in 6, the square root ends in a 4 or a 6.
- If the number ends in 5, the square root ends in a 5.
- If the number ends in 0, the square root ends in a 0.

For example, **√1849**, the number ends in 9. This means the last digit of our answer must be a 3 or a 7.

Step 2: Find the First Digit

Now, let's find the first digit of the answer.

- First, ignore the last two digits of the number. For $\sqrt{1849}$, we will ignore the 49.
- You are left with 18.
- Think of the perfect squares you know: $1^2=1$, $2^2=4$, $3^2=9$, $4^2=16$, $5^2=25$.
- Which of these perfect squares is the closest to 18, without going over? The answer is 16 (4^2).
- The square root of 16 is 4. This means the first digit of your answer is 4.

Putting It All Together

Now you know the answer starts with 4 and ends with either a 3 or a 7. Your possible answers are 43 or 47.

To figure out which one is right, simply multiply the first digit you found (4) by the next number (5):

$4 \times 5 = 20$

Now, compare that result (20) to the number we were working with in Step 2 (18).

- Since 18 is less than 20, you choose the smaller of the two possible last digits (3 or 7).
- The smaller one is 3, so the answer is 43!

If the number were, for example, $\sqrt{2209}$, the first digit would be 4, and you'd be choosing between 43 and 47. The number you'd compare to would be 22. Since 22 is greater than 20, the answer would be 47.

Try It Out!

Use this trick to find the square roots of these perfect squares:

- $\sqrt{676}$=?
- $\sqrt{1225}$=?
- $\sqrt{2916}$=?
- $\sqrt{4096}$=?
- $\sqrt{8649}$=?
- $\sqrt{1296}$=?
- $\sqrt{7744}$=?
- $\sqrt{9604}$=?

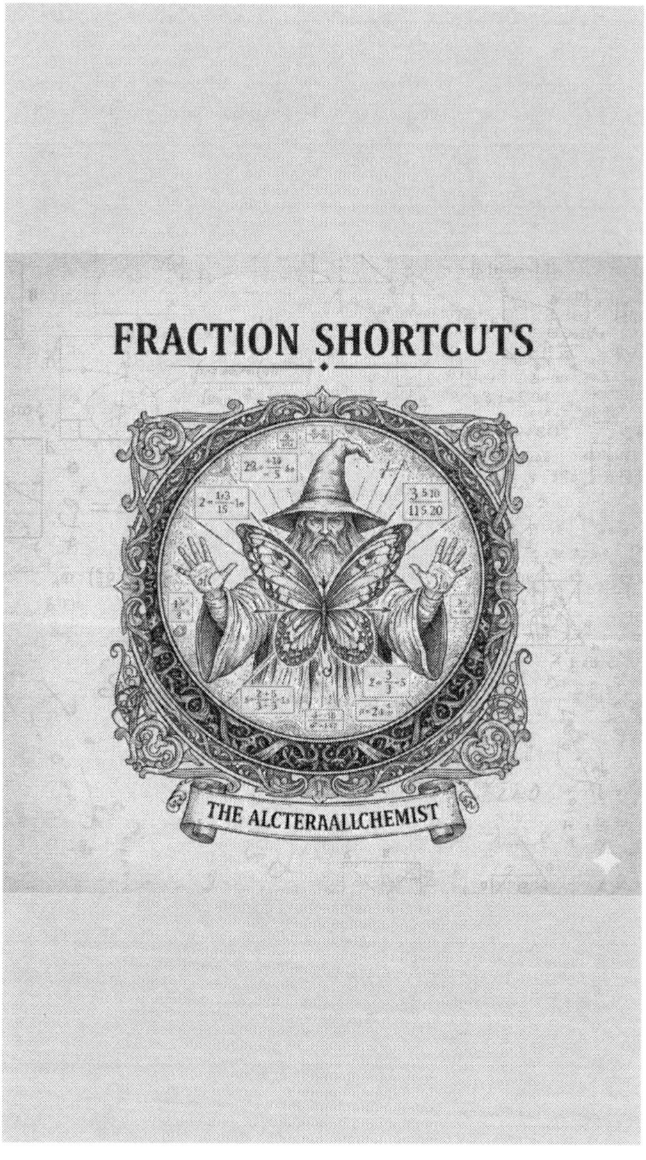
FRACTION SHORTCUTS
THE ALCTERAALLCHEMIST

Chapter- 9

Fraction Shortcuts

Fractions often get a bad reputation. The idea of adding, subtracting, or comparing them can seem messy and confusing. But with these simple tricks, you'll see that fractions are nothing to be afraid of. You can tame them and solve problems quickly and confidently.

The "Butterfly" Method for Comparing Fractions

Have you ever wondered which of two fractions is bigger, like or 75? It's not immediately obvious. The "Butterfly" method makes this comparison super easy, without needing a common denominator.

Let's compare 2/3and 4/5.

- **Step 1:** Draw a butterfly! Multiply the numerator of the first fraction by the denominator of the second. 2×5=10. Write this number above the first fraction.
- **Step 2:** Draw the other wing! Multiply the numerator of the second fraction by the denominator of the first. 3×4=12. Write this number above the second fraction.
- **Step 3:** Compare the two numbers you wrote on top.
    - Since 12 is greater than 10, it means that the fraction it came from (4/5) is the bigger fraction.

So, 4/5>2\3. This trick works every time, whether you're comparing two positive fractions or even negative ones.

Quick Addition and Subtraction

Adding and subtracting fractions with different denominators can be a pain. But there's a quick way to get the job done, similar to our butterfly method.

Let's add 1/2+1/3.

- **Step 1:** Multiply the denominators to get your common denominator: $2 \times 3 = 6$. This will be the bottom number of your answer.
- **Step 2:** Multiply diagonally (like the butterfly method). The first multiplication is $1 \times 3 = 3$. The second is $1 \times 2 = 2$.
- **Step 3:** Add the numbers you got in Step 2: $3+2=5$. This will be the top number of your answer.
- **Step 4:** Put it all together. The answer is 5/6.

You can use the same method for subtraction. For 1/2−1/3, the steps are the same, but you subtract the numbers you get in Step 2: $3-2=1$.

The answer is **1/6**.

Try It Out!

Use these shortcuts to solve the following problems in your head:

- Which fraction is bigger: 4/3 or 9/7 ?

- Which fraction is smaller: 9/3 or 5/8 ?

- Solve: 1/4+2/5=?

- Solve: 6/5−1/2=?

- Solve: $4/5-7/6=?$

- Solve: $9/2+2/7=?$

YOUR JOURNEY CONTINUES

Chapter- 10

Your Journey Continues

Congratulations! You've made it to the end of your secret training manual. By now, you've unlocked a set of powerful shortcuts that can change the way you think about numbers. You've seen that math isn't just about following a long, complicated list of steps. It's about being clever, looking for patterns, and finding the easiest way to get to the answer.

The tricks in this book are just the beginning. The most exciting part of this journey is that now you have the tools to discover your own shortcuts. As you continue to practice, you'll start to see patterns everywhere. You'll find your own unique ways to solve problems and you'll feel a sense of confidence and accomplishment every time you do.

Don't let these skills get rusty. The best way to keep your new superpowers is to use them! Challenge yourself with a few problems every day. Do some quick mental math when you're at the grocery store or calculating a tip at a restaurant.

Remember, a genius isn't just someone who knows all the answers—it's someone who knows how to think. You've just learned to think smarter.

Now, go out and amaze yourself with what you can do!

Appendix: Practice Problems

Here are some problems to help you practice the tricks you've learned. Try to solve them using the shortcuts you've mastered, without a calculator! The answers are at the very bottom.........

Addition and Subtraction

- 67+19=?

- 84+31=?

- 91−47=?

- $1000 - 345 = ?$

Multiplication and Division

- 32×11=?

- 45×5=?

- 12×9=?

- $96 \times 93 = ?$

Squares and Roots

- $35^2 = ?$

- $65^2 = ?$

- $\sqrt{441}$=?

- $\sqrt{961}$=?

Answers

- $67+19=86$
- $84+31=115$
- $91-47=44$
- $1000-345=655$
- $32\times11=352$
- $45\times5=225$
- $12\times9=108$
- $96\times93=8928$
- $35^2=1225$
- $65^2=4225$
- $\sqrt{441}=21$
- $\sqrt{961}=31$

About Author

Himanshri Chaudhary is proud founder of Himanshri Success School. She holds an M.Sc. in Mathematics and a Bachelor of Education (B.Ed.) degree. She has demonstrated her academic excellence and professional competence by qualifying for both the **CTET (Primary level)** and the **UPSSSC PET** exam. With a strong academic foundation and a deep-seated passion for teaching, she is committed to helping students and educators simplify and master mathematical calculations.

Drawing from her diverse experience and expertise, Himanshri has created this book to share practical short tricks and techniques designed to enhance speed, accuracy, and confidence in mathematics. Having successfully navigated competitive examinations herself, her goal is to make math accessible and enjoyable, supporting competitive exam aspirants, students, and teachers through clear and effective methods.

Beyond her professional pursuits, Himanshri enjoys nurturing young minds and exploring innovative approaches to mathematics education to bridge the gap between complex theory and practical application.

Made in the USA
Monee, IL
12 November 2026

7e80edf0-b321-46c9-841f-07cc027d746fR01